Proofs of Mathematical Problems

Book 2

Xu Feng

Alright, everyone, let me go on to show my Proofs of Mathematical Problems (The Book 2) to you all. In the Book 2, it be included: 1) On the Beal Conjecture, 2) On the Fermat Conjecture, 3) On the Hodge Conjecture, 4) On the Quintic Equation, 5) On the Perfect Cuboids, and 6) On $ax^2 + by^2 = z^2$.

These proofs of mathematical problems are the gifts to my mam Anhua Huang. I thank her for my grown life. Now, I still remember that in the past time, she taught me something about the truth: Don't give up what you want and do. So that I'm the lucky one. I also thank all my mathematical teachers when I had been educated. Especially in junior middle school, the main teacher in my class taught me the mathematics, besides which the other great points are that I can learn anything by myself and put the Chinese characters to the fucking hell. Of course, I don't write any book in Chinese, and don't read any Chinese book in my lifetime.

Also, I can feel that: so lonely but happy, this is me. I'll die happy for my Proofs of Mathematical Problems. But for God's sake, I don't have any time to write the novels now.

Yes, I'm a Dianbaian, not a Chinese. And China is not my country. Don't write anything about me in Chinese, because I hate that.

Then, I have something that I have to say: all my hometown(Dianbai)'s young children and people, you have to put the Chinese characters to the hell, and you have to learn English or other languages, so that you have to use English or other languages to think, write and speak. Don't write anything in Chinese anymore. The Chinese characters are shit like the China's culture. It just teaches you nothing. But unfortunately of the truth, the Chinese people have not understood that. They always think the China's culture is great. But when you know that especially the China's history books are full of the fucking fiction of the facts, you'll be shamed for them. But, in fact, the Chinese people don't. They love to talk about the fucking fiction history.

Also, of course, I don't understand why Canton's and Hong Kong's people use the fucking Chinese characters to write. They have Jyutping. So, when they use the Jyutping to speak, think and write, they are great.

In the future, Guangdong be an independent country.

But, if you love to do the mathematical problems, my suggestions are:

1) You have to know: what's the problem?
2) So what do you think? That's the most important, now using all of you learned, and forget the others said. Why? Because the others said like the fucking shit.
3) Then do it, and you solve it.

Problem 1: On the Beal Conjecture

I must say that Mr Beal is a great man, and his theorem is one of the greatest theorems all the times. So, now just cheers for you and me. And its great point is the common prime factor.

The equation be:

$$A^x + B^y = C^z \ .$$

Let $\ C^z = 2Q \ ,$

so that $\ A^x \ , \ Q = \dfrac{C^z}{2}$

and $\ B^y \ $ are the arithmetic progression,

its common difference is $\ d \ ,$

and it has two equations set:

$$\begin{cases} A^x + d = Q = \dfrac{C^z}{2} \\ A^x + 2d = B^y \end{cases}$$

But now, let $\ d = 0 \ ,$

so, $\ A^x = B^y = Q = \dfrac{C^z}{2} \ .$

Because $\ Q = \dfrac{C^z}{2} \ ,$

then it is $\ 2Q = C^z \ ,$

and $\ \log_C(2Q) = \log_C(2) + \log_C(Q) = z \ .$

also, because $\ \log_C(2)$

and $\ \log_C(Q) \ $ are the integers,

and $\ \log_C(2) > 0 \ , \ \log_C(Q) > 0 \ ;$

so, just only let $\log_C(2)=1$,

and $C=2$.

Then it has:

$$A^x=B^y=Q=\frac{2^z}{2}=2^{z-1}$$

Because A , B

and $C=2$ have a common prime factor 2,

that it has two parts.

Part 1) $A=2^{n+1}$, $B=2^n$;

Part 2) $A=2^n$, $B=2^{n+1}$.

Because $A^x=B^y$,

so $A=2^{n+1}$, $B=2^n$;

or $A=2^n$, $B=2^{n+1}$.

But now that it has :

part 1) $\left(2^{n+1}\right)^x=\left(2^n\right)^y$,

and $x=n$, $y=n+1$;

part 2) $\left(2^n\right)^x=\left(2^{n+1}\right)^y$,

and $x=n+1$, $y=n$.

Because $A^x=B^y=2^{z-1}$,

when $A=2^{n+1}$, $B=2^n$,

$$x=n \text{ , and } y=n+1 \text{ ,}$$

so it has :
$$z=(n+1)n+1 \text{ ,}$$

but now $x>2$, $y>2$,

and $z>2$,

that it means:

$$n=3,4,\dots,\infty \text{ .}$$

and the equation is :

$$\left(2^{n+1}\right)^{n}+\left(2^{n}\right)^{n+1}=2^{(n+1)n+1} \text{ ,}$$

$$n=3,4,\dots,\infty \text{ .}$$

Because $1=\log_2(2)$,

so, the equation is:

$$\left(2^{n+1}\right)^{n}+\left(2^{n}\right)^{n+1}=2^{(n+1)n+\log_2(2)} \text{ ,}$$

$$n=3,4,\dots,\infty \text{ ;}$$

$$x=n \text{ , } y=n+1 \text{ ,}$$

$$z=(n+1)n+\log_2(2) \text{ ;}$$

and in which $A=2^{n+1}$,

$$B=2^{n} \text{ and } C=2 \text{ are the integers.}$$

But now,

let the base number 2 to be any prime number (P),

and $P>1$,

the equation be :

$$\left(P^{n+1}\right)^{n}+\left(p^{n}\right)^{n+1}=P^{(n+1)n+\log_P(2)} \quad ,$$

$$n=3,4,\dots,\infty \quad ,$$

P = any prime number,

and $P>1$,

$$x=n \ , \quad y=n+1 \ ,$$

$$z=(n+1)n+\log_p(2) \ ;$$

so, $A=P^{n+1}$, $B=P^{n}$

and $C=P$ are the integers.

P.S. $\quad P^{(n+1)n+\log_P(2)}=P^{(n+1)n} \ P^{\log_P(2)} \quad ,$

and $P^{\log_p(2)}=2$,

$$P^{(n+1)n+\log_P(2)}=2P^{(n+1)n} \quad .$$

But when $A=P^{n}$, $B=P^{n+1}$,

the equation be :

$$\left(P^{n}\right)^{n+1}+\left(P^{n+1}\right)^{n}=P^{(n+1)n+\log_P(2)} \quad ,$$

$$x=n+1 \ , \quad y=n \ ,$$

$$z=(n+1)n+\log_P(2) \ ,$$

P = any prime number,

and $P>1$, $n=3,4,\dots,\infty$,

so, $A=P^n$, $B=P^{n+1}$

and $C=p$ are the integers.

Problem 2: On the Fermat Conjecture

Mr Fermat, you were wrong on $x^n+y^n=z^n$. So let me show my proof to you all.

The equation be :

$$x^n+y^n=z^n ,$$

let $z^n=2Q$, so x^n ,

$$Q=\frac{z^n}{2} \text{ and } y^n \text{ are the arithmetic progression,}$$

its common difference is d ,

and it has two equations set :

$$\begin{cases} x^n+d=Q=\dfrac{z^n}{2} \\ x^n+2d=y^n \end{cases} ,$$

Because $Q=\dfrac{z^n}{2}$,

so $2Q=z^n$

and $\log_z(2Q)=n$,

then it be $\log_z(2)+\log_z(Q)=n$.

Because $\log_z(2)$ and $\log_z(Q)$ are the integers ,

and $\log_z(2)>0$, $\log_z(Q)>0$;

but now only let $\log_z(2)=1$,

so that it be : $z=2$,

and $Q=\dfrac{2^n}{2}=2^{n-1}$.

Because $x^n+d=Q$,

and let $d=x^n$,

so, $x^n+x^n=2^{n-1}$,

and $x^n=2^{n-2}=\dfrac{2^n}{4}=\dfrac{1}{4}2^n$.

Then just let it be:

$$x^n=\frac{1}{4}2^n=b \quad,$$

so, $\log_x(b)=\log_x(\dfrac{1}{4}2^n)=n$;

because the logarithm be:

$$\log_x(b)=\frac{\log_c(b)}{\log_c(x)} \quad,$$

and let $c=2$,

that it means :

$$\log_x(b)=\frac{\log_2(b)}{\log_2(x)}=\frac{\log_2(\dfrac{1}{4}2^n)}{\log_2(x)}=n \quad,$$

by the next step, it is :

$$\frac{\log_2(\frac{1}{4})+\log_2(2^n)}{\log_2(x)}=n \quad,$$

then , it is: $\quad \dfrac{\log_2(\frac{1}{4})+n}{n}=\log_2(x) \quad,$

but now, it has : $\quad x=2^{\frac{\log_2(\frac{1}{4})+n}{n}} \quad.$

Because $\quad x^n+2d=y^n \quad,$

and $\quad d=x^n=\dfrac{1}{4}2^n \quad,$

so, $\quad 3x^n=y^n=\dfrac{3}{4}2^n \quad;$

then let $\quad 3x^n=y^n=g=\dfrac{3}{4}2^n \quad,$

that it means:

$$\log_y(g)=\log_y(\frac{3}{4}2^n)=n \quad.$$

Because the logarithm be:

$$\log_y(g) = \frac{\log_c(g)}{\log_c(y)} \quad ,$$

and let $c = 2$,

so, $\log_y(g) = \dfrac{\log_2(\frac{3}{4} 2^n)}{\log_2(y)} = \dfrac{\log_2(\frac{3}{4}) + \log_2(2^n)}{\log_2(y)} = n \quad ,$

then it has : $\log_2(y) = \dfrac{\log_2(\frac{3}{4}) + n}{n} \quad ,$

and $y = 2^{\frac{\log_2(\frac{3}{4}) + n}{n}}$.

So the equation is :

$$\left(2^{\frac{\log_2(\frac{1}{4}) + n}{n}}\right)^n + \left(2^{\frac{\log_2(\frac{3}{4}) + n}{n}}\right)^n = 2^n \quad ,$$

$$n = 3, 4, \ldots, \infty \quad ,$$

$$x = 2^{\frac{\log_2(\frac{1}{4}) + n}{n}}$$

and $y = 2^{\dfrac{\log_2(\frac{3}{4}) + n}{n}}$ are the irrationals ,

$z = 2$ is an integer.

But now, let the base number 2 to be any prime number (P),

and $P > 1$,

the equation be:

$$\left(P^{\dfrac{\log_p(\frac{1}{4}) + n}{n}} \right)^n + \left(P^{\dfrac{\log_P(\frac{3}{4}) + n}{n}} \right)^n = P^n ,$$

$n = 3, 4, \ldots, \infty$,

P = any prime number , and $P > 1$,

so, $x = P^{\dfrac{\log_P(\frac{1}{4}) + n}{n}}$

and $y = P^{\dfrac{\log_p(\frac{3}{4}) + n}{n}}$ are the irrationals,

$z = P$ are the integers.

So That the equation:

$$x^n + y^n = z^n \quad \text{has an integer (} \quad z = P \text{)},$$

when $n = 3, 4, \ldots, \infty$,

but now I call it Anhua-Diophantus theorem.

Problem 3: On the Hodge Conjecture

The Newton binomial theorem is a good thing, it can teach me something about the key of the Hodge conjecture.

Let $H_{2n} = (x+1)^{2n}$, $H_n = (x+1)^n$,

and $H_{2n}^{2k} = H_n^k + H_n^k$, $n \in N$,

so, $H_{2n} = (x+1)^{2n} = C_{2n}^0 x^{2n} + C_{2n}^1 x^{2n-1} + \ldots + C_{2n}^{2n-1} x + C_{2n}^{2n} 1$,

and $H_{2n}^{2k} = C_{2n}^{2k} x^{2n-2k}$.

$H_n = (x+1)^n = C_n^0 x^n + C_n^1 x^{n-1} + \ldots + C_n^{n-1} x + C_n^n 1$,

and $H_n^k = C_n^k x^{n-k}$.

Because $H_{2n}^{2k} = H_n^k + H_n^k$,

that it is : $C_{2n}^{2k} x^{2n-2k} = C_n^k x^{n-k} + C_n^k x^{n-k}$.

but now, let $\quad C_{2n}^{2k}=C_n^k \quad$,

that it be: $\quad x^{2n-2k}=x^{n-k}+x^{n-k} \quad$,

$$\frac{x^{2n}}{x^{2k}}=\frac{x^n}{x^k}+\frac{x^n}{x^k} \quad,$$

$$x^{2n}=x^{2k}\left(\frac{x^n}{x^k}+\frac{x^n}{x^k}\right) \quad.$$

Because $\quad \begin{cases} x^{2n}=x^n x^n \\ x^{2k}=x^k x^k \end{cases}$,

so, $\qquad x^{2n}=x^k\left(x^n+x^n\right) \quad$,

$$x^n=x^k+x^k=2x^k \quad,$$

then, let $\quad x^n=2x^k=b \quad$,

that it has: $\quad x^k=\dfrac{b}{2} \quad$, and $\quad \log_x\left(\dfrac{b}{2}\right)=k \quad$,

$$\log_x\left(\frac{b}{2}\right)=\log_x(b)-\log_x(2)=k \quad.$$

Because $\quad \log_x(b) \quad$ and $\quad \log_x(2) \quad$ are the integers,

and $\quad \log_x(b)>0 \quad$, $\quad \log_x(2)>0 \quad$,

so , let $\log_x(2)=1$,

that it has, $x=2$,

$2^n=2(2^k)$, and $n=k+1$.

now its equation is:

$$C_{2n}^{2k}2^{2n-2k}=C_n^k2^{n-k}+C_n^k2^{n-k} \; ,$$

$$n=k+1 \; , \text{ and } \; C_{2n}^{2k}=C_n^k$$

But, let the base number 2 to be any prime number (P),

$$P>1 \; ,$$

and $n=k+1$, $C_{2n}^{2k}=C_n^k$,

that it has these equations :

$$C_{2n}^{2k}3^{2n-2k}=C_n^k3^{n-k}+C_n^k3^{n-k}+C_n^k3^{n-k} \; ,$$

$$H_{2n}^{2k}=H_n^k+H_n^k+H_n^k \; ;$$

$$C_{2n}^{2k}5^{2n-2k}=C_n^k5^{n-k}+C_n^k5^{n-k}+C_n^k5^{n-k}+C_n^k5^{n-k}+C_n^k5^{n-k} \; ,$$

$$H_{2n}^{2k}=H_n^k+H_n^K+H_n^k+H_n^k+H_n^k \; ;$$

$$C_{2n}^{2k}7^{2n-2k}=C_n^k7^{n-k}+C_n^k7^{n-k}+C_n^k7^{n-k}+C_n^k7^{n-k}+C_n^k7^{n-k}+C_n^k7^{n-k}+C_n^k7^{n-k}$$

,

$$H_{2n}^{2k}=H_n^k+H_n^k+H_n^k+H_n^k+H_n^k+H_n^k+H_n^k \; ;$$

$$C_{2n}^{2k} P^{2n-2k} = C_n^k P^{n-k} + C_n^k P^{n-k} + ... + C_n^k P^{n-k} = P\left(C_n^k P^{n-k}\right) ,$$

$$H_{2n}^{2k} = H_n^k + H_n^k + ... + H_n^k = PH_n^k ,$$

P = any prime number,

and $P > 1$,

$$n = k+1 , \quad C_{2n}^{2k} = C_n^k$$

$$H_{2n}^{2k} = C_{2n}^{2k} P^{2n-2k} , \quad H_n^k = C_n^k P^{n-k} .$$

Problem 4: On the Quintic Equation

I'm so easy to do this problem.

The quintic equation be:

$$ax^5 + bx^4 + cx^3 + dx^2 + ex + f = 0 ,$$

$a, b, c, d, e,$ and $f \in R , a \neq 0 ,$

but in which it can be two equations set:

$$\begin{cases} ax^5 + bx^4 + cx^3 = 0 \\ dx^2 + ex + f = 0 \end{cases} ,$$

Part 1:

Because $ax^5 + bx^4 + cx^3 = 0$,

so, $x^3\left(ax^2 + bx + c\right) = 0$,

and forget x^3 ,

then just let $ax^2 + bx + c = 0$,

so, $\quad x^2 + \dfrac{b}{a}x + \dfrac{c}{a} = 0$,

$$\left(x + \dfrac{b}{2a}\right)^2 - \dfrac{b^2}{4(a)^2} + \dfrac{c}{a} = 0 \quad ,$$

$$\left(x + \dfrac{b}{2a}\right)^2 = \dfrac{b^2}{4(a)^2} - \dfrac{c}{a} = \dfrac{b^2 - 4ac}{4(a)^2} \quad ,$$

$$x + \dfrac{b}{2a} = \pm \dfrac{\sqrt{b^2 - 4ac}}{2a} \quad ,$$

$$x = \pm \dfrac{\sqrt{b^2 - 4ac}}{2a} - \dfrac{b}{2a} \quad ,$$

but now, let $\quad b^2 - 4ac = 0$,

so, $\quad b^2 = 4ac$, $\quad ac > 0$, and $\quad x = \pm \dfrac{b}{2a}$.

Now, because $\quad dx^2 + ex + f = 0$,

and $\quad x = \dfrac{b}{2a}$, or $\quad x = -\dfrac{b}{2a}$,

so that it has two equations set :

$$\begin{cases} d\left(\dfrac{b}{2a}\right)^2 + e\left(\dfrac{b}{2a}\right) + f = 0 \\ d\left(-\dfrac{b}{2a}\right)^2 + e\left(-\dfrac{b}{2a}\right) + f = 0 \end{cases} \quad ,$$

by the next step, in which $e = 0$,

and $\qquad d\left(\dfrac{b^2}{4(a)^2}\right) = -f$,

but $\qquad b^2 = 4ac$,

so $\qquad d\left(\dfrac{c}{a}\right) = -f$.

Now, when $e = 0$, $b^2 = 4ac$, $ac > 0$,

and $d\left(\dfrac{c}{a}\right) = -f$,

the quintic equation:

$$ax^5 + bx^4 + cx^3 + dx^2 + ex + f = 0$$

has a common real solution:

$$x = \dfrac{b}{2a} \quad ,$$

or $\qquad x = -\dfrac{b}{2a}$.

I.E. let $a=1$, $c=4$, $b=-4$,

$d=2$, $f=-8$, and $e=0$,

the quintic equation :

$$x^5-4x^4+4x^3+2x^2-8=0 \quad ,$$

it has a common real solution:

$$x=2 \quad , \text{ or } \quad x=-2 \quad .$$

Part 2 :

The quintic equation to be two equations set :

$$\begin{cases} ax^5+bx^4+cx^3=0 \\ dx^2+ex+f=0 \end{cases} ,$$

then, $x^3(ax^2+bx+c)=0$,

and forget x^3 ,

so, $ax^2+bx+c=0$.

Because $dx^2+ex+f=0$,

$$x^2+\frac{e}{d}x+\frac{f}{d}=0 \quad ,$$

$$\left(x+\frac{e}{2d}\right)^2-\frac{e^2}{4(d)^2}+\frac{f}{d}=0 \quad ,$$

$$x+\frac{e}{2d}=\pm\frac{\sqrt{e^2-4df}}{2d} \quad ,$$

$$x=\pm\frac{\sqrt{e^2-4df}}{2d}-\frac{e}{2d} \quad ,$$

but now let $e^2 - 4df = 0$,

and $df > 0$,

so, $x = \pm \dfrac{e}{2d}$.

because $ax^2 + bx + c = 0$,

it has two equations set :

$$\begin{cases} a\left(\dfrac{e}{2d}\right)^2 + b\left(\dfrac{e}{2d}\right) + c = 0 \\ a\left(-\dfrac{e}{2d}\right)^2 + b\left(-\dfrac{e}{2d}\right) + c = 0 \end{cases} ,$$

and $e^2 = 4df$,

then $b = 0$, $\dfrac{af}{d} = -c$.

So, when $b = 0$, $e^2 = 4df$,

$df > 0$, and $\dfrac{af}{d} = -c$,

the quintic equation :

$$ax^5 + bx^4 + cx^3 + dx^2 + ex + f = 0$$

has a common real solution :

$$x = \dfrac{e}{2d} \text{ , or } x = -\dfrac{e}{2d} .$$

I.E. let $a = 2$, $b = 0$, $c = -8$,

$d = 1$, $e = -4$, and $f = 4$,

the quintic equation :

$$2x^5 - 8x^3 + x^2 - 4x + 4 = 0$$

has a common real solution:

$$x=2 \quad , \text{or} \quad x=-2 \quad .$$

Problem 5: On the Perfect Cuboids

I'm so happy to do these fucking great equations.

The equations are :

$$\begin{cases} a^2+b^2=d^2 & \text{①} \\ b^2+c^2=e^2 & \text{②} \\ c^2+a^2=f^2 & \text{③} \\ a^2+b^2+c^2=g^2 & \text{④} \end{cases}$$

① $a^2+b^2=d^2$,

let $d^2=2Q$, and $d=2^n$, $n=1,2,...,\infty$,

so, a^2 , $Q=\dfrac{d^2}{2}=\dfrac{(2^n)^2}{2}$, and b^2 are

the arithmetic progression, and its common difference is k ,

① it has these equations set :

$$\begin{cases} a^2+k=Q=\dfrac{d^2}{2}=\dfrac{(2^n)^2}{2} \\ a^2+2k=b^2 \end{cases} ,$$

now let $k=a^2$,

that it has $\quad 2a^2 = \dfrac{(2^n)^2}{2}$,

$$2a = 2^n \ ,$$

$$a = 2^{n-1} \ ,$$

and $\quad 3a^2 = b^2$,

$$3(2^{n-1})^2 = b^2 \ ,$$

$$\sqrt{3}(2^{n-1}) = b \ .$$

② $\quad b^2 + c^2 = e^2$,

let $\quad e^2 = 2J$,

so, $\quad b^2$, $\quad J = \dfrac{e^2}{2}$

and $\quad c^2 \quad$ are the arithmetic progression,

and its common difference is $\quad m$,

that it has two equations set :

$$\begin{cases} b^2 + m = J = \dfrac{e^2}{2} \ , \\ b^2 + 2m = c^2 \end{cases}$$

because $\quad b = \sqrt{3}(2^{n-1})$,

so, the two equations are :

$$\begin{cases} 3(2^{n-1})^2 + m = \dfrac{e^2}{2} \\ 3(2^{n-1})^2 + 2m = c^2 \end{cases},$$

for e are the integers,

only let $m = \dfrac{3}{2}(2^{n-1})^2$,

so, $3(2^{n-1})^2 + \dfrac{3}{2}(2^{n-1})^2 = \dfrac{e^2}{2}$,

$$9(2^{n-1})^2 = e^2 ,$$

$$e = 3(2^{n-1}) ,$$

and $3(2^{n-1})^2 + 2 \times \dfrac{3}{2}(2^{n-1})^2 = c^2$,

$$6(2^{n-1})^2 = c^2 ,$$

$$c = \sqrt{6}(2^{n-1}) .$$

③ $c^2 + a^2 = f^2$,

because $c^2 = 6(2^{n-1})^2$,

$$a = (2^{n-1})^2 ,$$

so, $6(2^{n-1})^2 + (2^{n-1})^2 = f^2$,

$$f = \sqrt{7}(2^{n-1}) .$$

④ $a^2 + b^2 + c^2 = g^2$,

because $a^2 = (2^{n-1})^2$,

$$b^2 = 3\left(2^{n-1}\right)^2 ,$$

and $\quad c^2 = 6\left(2^{n-1}\right)^2 ,$

so, $\quad \left(2^{n-1}\right)^2 + 3\left(2^{n-1}\right)^2 + 6\left(2^{n-1}\right)^2 = g^2 ,$

$$g = \sqrt{10}\left(2^{n-1}\right) .$$

Now, in the cuboids,

$$a = \left(2^{n-1}\right) , \quad d = 2^n \quad \text{and} \quad e = 3\left(2^{n-1}\right)$$

are the integers;

$$b = \sqrt{3}\left(2^{n-1}\right) , \quad c = \sqrt{6}\left(2^{n-1}\right) ,$$

$$f = \sqrt{7}\left(2^{n-1}\right) \quad \text{and} \quad g = \sqrt{10}\left(2^{n-1}\right)$$

are the irrationals.

Problem 6: On $\quad ax^2 + by^2 = z^2$

The Problem is so easy when I do it.

The equation be :

$$ax^2 + by^2 = z^2 ,$$

let $\quad z^2 = 2Q ,$

so, ax^2 , $\quad Q = \dfrac{z^2}{2} \quad$ and $\quad by^2$

are the arithmetic progression,

and its common difference is $\quad d$,

that it has two equations set:

$$\begin{cases} ax^2 + d = Q = \dfrac{z^2}{2} \\ ax^2 + 2d = by^2 \end{cases},$$

but now, let $d = ax^2$,

so, the equations are :

$$\begin{cases} ax^2 + ax^2 = Q = \dfrac{z^2}{2} \\ ax^2 + 2ax^2 = by^2 \end{cases},$$

then $x^2 = \dfrac{z^2}{4a}$,

$$\log_x\left(\dfrac{z^2}{4a}\right) = 2 \ ,$$

$$\log_x(z^2) - \log_x(4a) = 2 \ ,$$

because $\log_x(z^2)$

and $\log_x(4a)$

are the integers,

and $\log_x(z^2) > 0$,

$$\log_x(4a) > 0 \ ,$$

now let $x = 4a$, and $a > 0$,

so, $\log_x(z^2) = 3$,

$$x^3 = z^2 \ ,$$

$$z = 8a(\sqrt{a}) \ ,$$

because $3ax^2 = by^2$, $x = 4a$,

so, $y = 4a\left(\sqrt{\dfrac{3a}{b}}\right)$, and $\sqrt{\dfrac{3a}{b}} > 0$,

but now, let $\sqrt{\dfrac{3a}{b}} = 2$,

$$3a = 4b \ , \ y = 8a \ .$$

because $\sqrt{a} > 0$,

and let $\sqrt{a}$ are the integers,

so, $a = 2^{2n}$,

then, these have :

$$x = 4\left(2^{2n}\right) \ , \ b = \frac{3\left(2^{2n}\right)}{4} \ ,$$

$$y = 8\left(2^{2n}\right) \ , \ z = 8\left(2^{2n}\right)\left(2^{n}\right) \ .$$

and the equation is :

$$\left(2^{2n}\right)\left[4\left(2^{2n}\right)\right]^2 + \frac{3\left(2^{2n}\right)}{4}\left[8\left(2^{2n}\right)\right]^2 = \left[8\left(2^{2n}\right)\left(2^{n}\right)\right]^2 \ ,$$

$$x = 4\left(2^{2n}\right) \ , \ y = 8\left(2^{2n}\right) \ ,$$

and $z = 8\left(2^{2n}\right)\left(2^{n}\right)$ are the integers.

But now, let the base number 2 to be any prime number(P),

$$P > 1 \ ,$$

the equation be :

$$P^{2n}[4(P^{2n})]^2 + \frac{3(p^{2n})}{4}[8(P^{2n})]^2 = [8(p^2 2n)(P^n)]^2 \quad,$$

$$x = 4(P^{2n}) \quad, \quad y = 8(P^{2n}) \quad,$$

and $z = 8(P^{2n})(P^n)$ are the integers.

www.ingramcontent.com/pod-product-compliance
Lightning Source LLC
Chambersburg PA
CBHW080630180526
45168CB00007B/3118